Contents

1 Different Cells for Different Jobs

A cell is a remarkable thing. A single cell alone is too small to be seen without a microscope, and yet this tiny chemical package has all the properties of life. The plants and animals we see around us, ourselves included, are built from millions and millions of cells, all working together. Whether it be an oak tree or a daisy, a killer whale or a human child, all living things are made of cells. Cells are the units of life. They are life's building blocks. In this book we will be looking at how cells join together in plants and fungi.

Cell types

With optical microscopes, it is not possible to look in detail at the structure of a cell. But when powerful electron microscopes became available to biologists in the 1960s, they revealed some of the cell's most important secrets.

There are basically two different types of cells. The simplest cells are the bacteria. These fast-working, living chemical factories have a simple structure of a thin outer membrane and usually a strong cell wall around a complex mix of water and chemicals.

The plant kingdom includes Earth's oldest and biggest living things. Giant Sequoias (*Sequoia dendron giganteum*) are the biggest. A trunk can be 39 feet (12 meters) wide, and they can weigh more than 600 tons. They can also live for 3,000 years, although bristlecone pines (*Pinus aristata*) live longer.

PLANTS & FUNGI
MULTICELLED LIFE

Robert Snedden

Series Editor
Andrew Solway

Heinemann Library
Chicago, Illinois

© 2003 Reed Educational & Professional Publishing
Published by Heinemann Library,
an imprint of Reed Educational & Professional Publishing,
Chicago, Illinois

Customer Service 888-454-2279

Visit our website at www.heinemannlibrary.com

Designed by Paul Davies and Associates
Illustrations by Wooden Ark
Originated by Ambassador Litho Ltd.
Printed by Wing King Tong in Hong Kong

07 06 05 04
10 9 8 7 6 5 4 3 2

Library of Congress Cataloging-in-Publication Data
Snedden, Robert.
  Plants & fungi : multicelled life / Robert Snedden.
      v. cm. -- (Cells and life)
  Includes index.
  Contents: 1. Different cells for different jobs: plant cells -- 2. The plant body: simple tissues, complex tissues, the epidermis -- 3. Stems: woody stems -- 4. Looking at leaves: food factories, photosynthesis, leaf specializations -- 5. Roots -- 6. A new generation: sexual reproduction, flowers, pollination and fertilization, seeds and germination -- 7. Fungi: working together.
    ISBN 1-58810-675-6 (HC)   1-58810-937-2 (Pbk.)
    1. Botany--Anatomy--Juvenile literature [1.Botany--Anatomy]. 2. Plants--Juvenile literature. [2. Plants.] 3. Fungi--Anatomy--Juvenile literature [3. Fungi]. I. Title. II. Series.
  QK671 .S64 2002
  580--dc21

                                        2001008693

Acknowledgments
The author and publishers are grateful to the following for permission to reproduce copyright material:
pp. 4, 17, 23, 31, 34, 38 Photodisc; pp. 5, 8, 13, 14, 16, 19, 24, 25, 22, 27 Oxford Scientific Films; pp.7, 10, 11, 20, 32, 43 Science Photo Library; pp. 15, 22, 33 J. Burgess/Science Photo Library; p. 21 Garden & Wildlife Matters Photo Library; p. 28 Stone; p. 37 K. Kent/Science Photo Library; p. 39 D. Scharf; p. 41 A. and H-F Michler; p. 42 S. Fraser;

Cover photograph reproduced with permission of Science Photo Library/Andrew Syred.

Every effort has been made to contact copyright holders of any material reproduced in this book.
Any omissions will be rectified in subsequent printings if notice is given to the publisher.

Our thanks to Richard Fosbery for his comments in the preparation of this book, and also to Alexandra Clayton.

Some words are shown in bold, **like this.** You can find out what they mean by looking in the glossary.

Edible mushrooms are the best-known fungi. Ceps, or porcini mushrooms (*Boletus edulis*), are wild mushrooms found in woodlands. They are highly prized for their flavor.

All other cells, including all the other single-celled organisms and all the cells that make up plants and animals, are more complex. They are divided up inside into various compartments, called **organelles.** Each organelle carries out a variety of specialized tasks inside the cell. These more complex cells have another characteristic feature—the **nucleus.** This is a compartment containing the cell's genetic material, the information center that guides the cell's activity. (A few cells, such as **phloem** cells in plants and red blood cells, lose their nucleus when they mature.)

Cells with nuclei are called **eukaryotes,** which means "true nucleus." Cells without a nucleus, the bacteria, are called **prokaryotes,** which means "before the nucleus."

Becoming multicellular

Many millions of years ago, the first multicellular organisms may have appeared when some single-celled organisms grouped together and formed a colony. In the beginning, all the cells in the colony would have been the same. But gradually different groups of cells would have taken on different jobs. This was the beginning of multicellular life.

In multicelled plants and animals today, billions of cells work together to support life. Some supply food, some transport food and wastes, others support the structure, and others produce the next generation. Cells of the same type form **tissues,** and several different tissues are combined in an organ, such as an animal's heart or brain. Groups of organs make up systems, such as the nervous system in animals or the **vascular** (transport) **system** in plants.

Being multicellular requires cooperation on a large scale. All around us in the living world we see this cooperation working flawlessly, time after time.

Plant Cells

A typical plant cell and a typical animal cell are similar in many ways. Both are tiny chemical factories held together by an outer skin. The skin of the cell is a thin **cell membrane.** In addition to holding the cell together, the membrane controls what passes in and out of the cell. It lets in some substances, but not others. For this reason it is described as a partially permeable membrane.

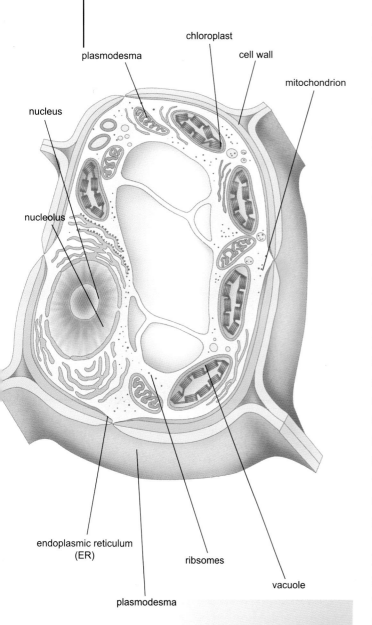

plasmodesma

chloroplast

cell wall

mitochondrion

nucleus

nucleolus

endoplasmic reticulum (ER)

ribsomes

vacuole

plasmodesma

This diagram shows the features of a plant cell. The main features of a plant cell not found in animal cells are the cell wall, plastids such as chloroplasts, and the vacuole.

Animal cells come in an enormous variety of shapes, but plant cells are usually boxlike or many-sided structures. This is because each plant cell is enclosed in a rigid cell wall made of cellulose, a tough material that gives plants their strength. Where the plant cell surface is exposed to air, waxes and other substances are produced to waterproof the cell. Inside the plant, the cell walls are sticky and cement the cells together.

Both plant and animal cells have a large **nucleus.** This is the cell's control center. It guides the activities of the cell by providing the instructions for building **proteins.** Proteins control the day-to-day activities of the cell. A cell is essentially a tiny chemical factory, with hundreds of different chemical reactions going on at any given moment. Special proteins called **enzymes** control this chemical activity by altering the rates of the various chemical reactions. Without enzymes, the reactions would virtually grind to a halt, and life would cease.

Between the nucleus and the cell membrane is the **cytoplasm.** This is where the cell obtains energy from food, carries out repairs, and makes new cell parts. The chemical reactions that go on in the cell are guided by instructions from the nucleus. Together, these reactions make up the cell's **metabolism.**

Plastids

In the cytoplasm of plant cells, there is one kind of structure not found in an animal cell. They are called plastids, and there are three types: **chloroplasts, chromoplasts,** and **amyloplasts.** The chloroplasts give plants their green color. They are full of a light-trapping pigment called **chlorophyll.** Inside the chloroplasts the energy of sunlight is converted into chemical energy, which is used to build sugars and other organic compounds.

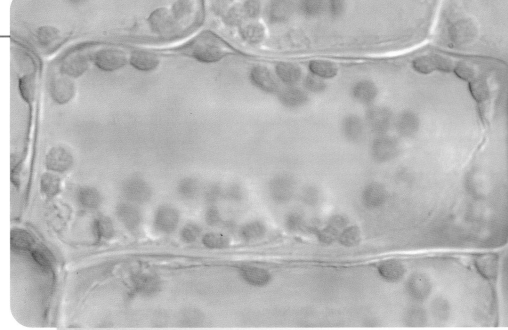

The oval green particles within these Canadian pondweed cells (*Elodea* species) are chloroplasts. Magnification approx. x 2,500.

Chromoplasts contain pigments of other colors. They give color to the plants' flower **petals,** ripening fruits, and autumn leaves.

Amyloplasts do not have pigments. They are where some plant cells store food reserves in the form of starch grains. Seeds, as well as tubers and root vegetables such as potatoes and carrots, have many amyloplasts.

Central vacuole

Another distinctive characteristic of a mature plant cell is the **vacuole.** This is a large central cavity filled with a watery fluid called cell sap. It is surrounded by a membrane similar to the cell membrane. The vacuole stores substances such as amino acids and sugars. These chemicals are vital for the cell's well-being. The vacuole is also used to store toxic wastes that would be harmful to the cell if they stayed in the cytoplasm.

The central vacuole can take up anywhere from 50 to 90 percent of the interior of the cell. It pushes the cytoplasm against the inside of the cell wall. This makes the cell turgid, or firm and plump. When a plant has sufficient water, all its cells are turgid and this keeps the plant upright. If fluid is lost from the vacuole, the cell collapses. This is what happens when a plant in need of water wilts.

2 The Plant Body

Plants are a large and diverse group of organisms, but they all have a common characteristic that separates them from animals. Plants can make their own food from very simple ingredients in the soil and air. They do this by capturing energy from sunlight and using it to build the complex organic **molecules** that are part of all living things. This process is known as **photosynthesis.** Because it can photosynthesize, a plant has no need to move around in search of food the way an animal does. All the energy a plant needs comes from the sun. This stationary lifestyle is reflected in the organization of a typical plant into two basic systems: shoots and roots.

A flowering plant

Angiosperms are so dominant in the plant world that most of the plant structures discussed in this book will be those of a typical flowering plant.

A flowering plant can be divided into two parts: the part above ground (the **shoot system**) and the part below ground (the **root system**).

flowers

stem

shoot system

leaf

root system

The shoot system can be further divided into leaves, stems, flowers, fruits, and other structures. Photosynthesis takes place inside a plant's leaves. The leaves are supported by the stem, which is also the main pipeline for water and **minerals** traveling from the roots to other parts of the plant. Flowers are the plant's reproductive organs. Fruits develop from flowers and are responsible for spreading a plant's seeds as widely as possible.

The root system spreads downward and outward through the soil. Roots absorb water and minerals from the soil. Minerals are vital for making the large, complex molecules needed for life. The roots also provide a firm anchorage for the shoot system above ground.

This is the shoot and root system of a cowslip (*Primula veris*).

The plant kingdom

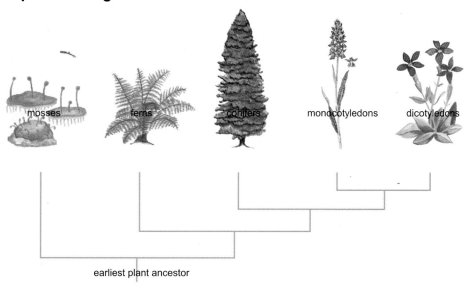

The vast majority of plants are flowering, seed-bearing plants (angiosperms). About 260,000 of the 300,000 or so known plant species are angiosperms. The angiosperms are divided into two major groups. The **dicotyledons,** or **dicots,** produce seeds that contain an embryo with two seed-leaves **(cotyledons).** They have broad leaves with branched veins. Most herbaceous plants, such as lettuce and daisies, are dicots. Flowering shrubs and trees and cacti also are dicots. The **monocotyledons (monocots)** have seeds containing a single seed-leaf. They have narrow leaves with straight parallel veins. Orchids, lilies, palms, and grasses such as rice, wheat, and corn, are all monocots.

- The **gymnosperms** (naked seeds) are shrubs or trees. The most abundant of them are **conifers** such as pine, fir, spruce, and cedar trees. They differ from angiosperms in having naked seeds. This means they do not form a fruit around the seed, as flowering plants do. There are about 750 species of gymnosperms.

- **Bryophytes** are the second-largest plant group. They include such plants as mosses, liverworts, and hornworts. These are low-growing plants with leaflike, stemlike, and rootlike parts, but they lack the complex **tissues** found in the flowering plants. There are about 19,000 species of bryophytes.

- Ferns, of which there are about 12,000 species, have roots and stems but do not flower. They produce spores on the undersides of their leaflike fronds.

3 Plant Tissues

As in all multicellular organisms, numerous cells of the same type form **tissues** in plants. All of the parts of a flowering plant, whether stem, leaf, or root, are made up of the following three tissue types:

- **Ground tissue** makes up the bulk of the plant. It consists of mainly simple, unspecialized cells. The **photosynthetic** cells in leaves are ground tissue, as are the pith (innermost layer) of stems and roots, and the soft tissues in fruits.

- The **vascular tissue** is the plant's plumbing. It has the task of transporting water and dissolved substances to all parts of the plant. The veins in a leaf are made up of vascular tissue.

- The **dermal tissue** is the plant's **epidermis,** or skin, which covers and protects the outer surfaces of the plant. It can be as thin as a single layer of cells, or more than three feet (one meter) thick, as in the bark of giant redwood trees.

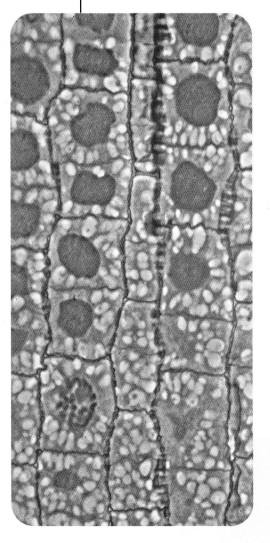

Meristems

Plants do not grow everywhere at the same time. Most of the growth in a plant happens in special areas called **meristems.** Meristem cells are small and have few **vacuoles.** They can divide continuously.

There are two main types of meristematic tissue.
- Apical meristems are found in the tips of roots, stems, and branches and in flower and leaf buds. Growth from apical meristems increases the length of the root or shoot. This is known as the plant's **primary growth.** Some of these cells will form other more specialized cells.
- Lateral meristem tissue is found just beneath the outer layer of roots and stems and is responsible for their thickening. This is known as **secondary growth.** One form of lateral meristem forms a sturdy covering that will replace the plant's epidermis.

This is a light micrograph of the apical meristem of an onion root. The blue spots in the cells are **nuclei.** Some of the cells seem to have two nuclei. This is because they are in the process of dividing. Magnification approx. x 320.

Simple tissues

The ground tissue of a plant is made up of three simple plant tissues: **parenchyma, collenchyma,** and **sclerenchyma.** Each of these tissues consists of just one type of cell.

Parenchyma

The simplest plant cells are the unspecialized cells called parenchyma. These cells are sometimes referred to as packing cells because they can fill spaces anywhere in the plant. Parenchyma cells have many sides and are roughly cube-shaped. They have thin walls and are easily pushed out of shape by the pressure of cells around them. Tissue composed of parenchyma cells is simply referred to as parenchyma. Most of the primary growth of stems, leaves, flowers, fruits, and roots comes from parenchyma cells.

More specialized cells in the plant lose their ability to divide as they mature, but parenchyma cells can divide throughout their lives. They are important for healing wounds in the plant. **Mesophyll** is a type of parenchyma that contains **chloroplasts** and is found in leaves. Other types of parenchyma are used for storage and as support tissues for the plant's **vascular system.**

Collenchyma and sclerenchyma

Collenchyma are elongated cells with unevenly thickened walls. Their role is to provide support. Collenchyma often form flexible ribs in leaf stalks and provide an effective strengthening system for young plant tissues. The strands on the outside of a celery stalk are made up of collenchyma cells.

Sclerenchyma cells have greatly thickened cell walls and provide rigid support for the plant. The thick walls of sclerenchyma are impregnated with a tough, waterproof material called **lignin.** Once the cell walls have thickened, the sclerenchyma cell itself dies.

Many seeds are also protected by sclerenchyma. The hard coat of a coconut shell and the stones of fruits such as cherries and peaches are formed by thick-walled sclerenchyma cells. Columns of long, thin sclerenchyma cells provide the structural strength to support stems and protect the cells within from drying out.

A light micrograph of a **dicot** plant stem in cross section shows parenchyma (yellow), collenchyma (grey outer ring), and sclerenchyma (grey inner ring) cells. Magnification approx. x 14.

Vascular Tissues

Complex **tissues** in a plant are those made up of more than one type of cell. Both the plant's **vascular system** and its **epidermis** are classified as complex tissues.

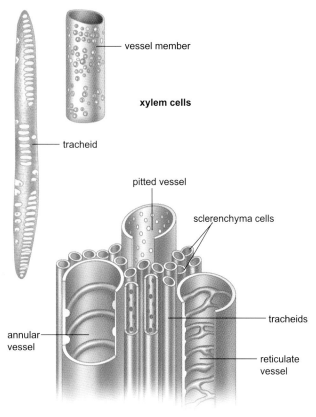

xylem cells

tracheid

pitted vessel

sclerenchyma cells

tracheids

annular vessel

reticulate vessel

xylem tissue

vessel member

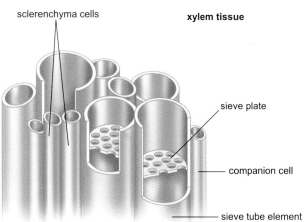

sclerenchyma cells

sieve plate

companion cell

sieve tube element

phloem tissue

These are xylem and phloem cells from a flowering plant.

The **vascular tissues** of a plant move water, dissolved **minerals,** and nutrients around the plant. The veins that you can see running through a plant's leaves are an obvious part of the vascular system. The vascular system is the equivalent of the circulatory system in animals. There are two types of tissues involved in the vascular system—**xylem** and **phloem.** Xylem and phloem are organized as a network of pipelines running through the plant.

Xylem

Xylem is the plant's water transport system, moving water and dissolved minerals through the plant. It also provides support. Xylem is mainly composed of two cell types, **tracheids** and **vessel members,** or vessel elements. Both types are long cells that join together in columns along shoots and roots. They have thick cell walls that are strengthened and waterproofed with **lignin.** Once xylem cells have reached full size, they die. The **cytoplasm** of the cell disappears, and all that is left is the cell wall.

Tracheids are long, narrow cells with tapering ends. The cell walls are dotted with tiny holes called pits, which allow water to flow in and out of the cells alongside the xylem. Vessel members are wider than tracheids and are found only in flowering plants. As vessel cells develop, their outer walls link up to form rigid pipelines through the plant. Where two cells meet, the end walls are at first perforated with holes and eventually disappear altogether. Vessel members conduct water much better than the narrow tracheids.

Phloem

Phloem carries sugars and other dissolved nutrients around the plant. Like xylem, it consists mainly of two types of cells. These are called **sieve tube members** and **companion cells.** Both have very thin cell walls, unlike the thick-walled xylem cells. Also unlike xylem cells, the phloem cells do not die when they are fully grown. This is because moving nutrients is an active process. Phloem cells must use energy to move sugars and other nutrients through the system.

Sieve tube members are the main cells through which nutrients are transported in the plant. The sieve cells are joined together at their ends, but unlike xylem tubes, the ends do not disappear. Instead, they contain a number of holes that line up with corresponding holes in the next cell. These perforated end walls are called **sieve plates.** Mature sieve cells lose their **nucleus.**

You might wonder how sieve cells get the instructions they need to work properly without a nucleus. Under the microscope, it can be seen that at least one companion cell is associated with each sieve cell. Companion cells keep their nucleus, and it is thought that one of the functions of the companion cell is to provide genetic information both for itself and its sieve cell sister.

The bark of this maritime pine tree (*Pinus pinaster*) has been tapped for its resinous sap, which will be turned into turpentine. The tree's phloem tissue, which carries the sap, is just beneath the bark.

The Epidermis

The outer surfaces of a plant are protected by a layer of cells called the **epidermis.** These unspecialized cells are tightly packed together and mostly they have no **chloroplasts.** Waxes and a fatty substance called cutin coat the outer surfaces of the cells, forming a protective coat called a **cuticle.**

This outermost layer of cells is very important to a plant's well-being. The cuticle helps to prevent water loss and also gives a degree of protection from attack by microorganisms. It is also transparent, which allows light to reach the photosynthetic **tissues** within the plant.

Stomata

The epidermis is not a continuous, unbroken sheet of cells. If it was, plants would suffocate and die. The surface of a plant's stems and leaves is dotted with tiny openings called **stomata.**

Plants, like most living things, require a continuous supply of oxygen. Oxygen is an essential ingredient for **respiration,** the process by which living things break down nutrients from food to get energy. During respiration, oxygen is taken up by the cells and carbon dioxide is produced as a waste product. Plants also need carbon dioxide from the air, which is used in **photosynthesis.** Photosynthesis uses up carbon dioxide and produces oxygen as a waste product.

In the hours of daylight, the rate of photosynthesis far exceeds that of respiration, so overall the plant takes in carbon dioxide and gives off oxygen. At night there is no photosynthesis, so the plant takes up oxygen and gives off carbon dioxide. This constant flow of gases takes place through the stomata.

There are thousands of stomata on every square inch of a leaf. In most plants the stomata are open during the day, when photosynthesis takes place, so that carbon dioxide can get in.

Plants such as geraniums (*Geranium* species) have hairs in the epidermal layer of cells that provide additional protection against insects.

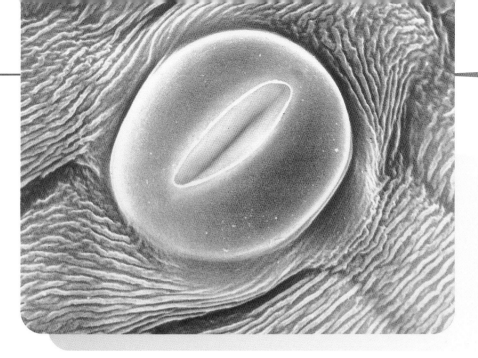

This is a single stoma. One leaf may have a million or more stomata, but they are so small, they make up less than one percent of the leaf's surface. Magnification approx x 1,000.

Around each stoma there are a pair of specialized cells called **guard cells.** When these cells take up water from the epidermal cells around them, they swell up. This makes them bend in such a way that an opening forms between them. When the guard cells lose water, they collapse against each other, closing the gap once more.

Guard cells are the only cells in the epidermis that contain chloroplasts. These chloroplasts are an important part of the mechanism by which the stomata open and close. During the hours of daylight, the chloroplasts in the guard cells photosynthesize and use up carbon dioxide. The low carbon dioxide levels trigger a process that brings water into the cells. The stoma then opens. At night photosynthesis stops, and the carbon dioxide levels in the guard cells rise again. This causes water to flow out of the guard cells, which collapse and close the stoma.

Transpiration

Most of the water a plant takes up through its roots is lost again through the stomata. The loss of moisture from the leaves and other plant parts is called **transpiration.** Transpiration actually provides the force that keeps water flowing up from the roots to the leaves. Water forms a continuous column as it flows from the roots, through the stem, and into the leaves. As water is lost through evaporation, the entire column of water is pulled upward, and more water is pulled in through the roots. Transpiration is strong enough to draw water to the top of a 328-foot (100-meter) tree.

4 The Stems of Plants

The stem is the part of a plant that produces and supports all of the above-ground parts of the plant: the buds, leaves, flowers, and fruit. The stem is the main pipeline that carries water and dissolved **minerals** from the roots and sugars manufactured in the leaves to other parts of the plant. In many plants, the stem ensures that the leaves spread to effectively gather sunlight for **photosynthesis.** A few kinds of stems grow underground or horizontally along the ground.

All the seed-bearing plants (the **angiosperms** and **gymnosperms**) have stems. Simpler plants such as liverworts, hornworts, and mosses do not. Plant stems vary greatly in size and appearance from one species to another. A cauliflower has a short, stubby stem, for example, but the trunk of a tree is a huge stem that in some species can be more than 328 feet (100 meters) long.

Buds

The plant stem produces buds from which new shoots, leaves, and flowers will grow. The bud is often protected by a cluster of modified leaves, called bud scales. Bud scales prevent water loss and protect the delicate growing **tissues** of the bud. Buds develop on the stem at points called nodes, on the top side of the angle where leaves attach to the stem. Many plants have a shoot tip, or terminal bud, at the end of each shoot. This growing point is an apical **meristem.** As the stem grows, plant tissues beneath the apical meristem gradually become more specialized and **differentiate** to form new leaves, flowers, and stems.

Herbaceous and woody stems

Herbaceous plants are those without woody stems, such as grasses and daisies. Herbaceous stems have soft tissue and grow very little in diameter. Most plants with herbaceous stems are annuals, living for only one growing season. Herbaceous stems consist mainly of primary tissues that develop from the apical meristem.

A tree's bark protects it from many types of injury, but some insects can cause serious damage. Bark beetles (family Scolytidae) bore chambers beneath the bark, where they lay their eggs. A bad infestation of these beetles can kill a tree. Magnification approx. x 23

Trees and shrubs have tough woody stems. Each growing season, woody stems develop new tissues that cause them to grow in diameter. The new layers form the annual rings visible in the trunk of a tree that has been cut down.

During their first year of growth, woody stems begin to develop secondary tissues. These support or replace primary tissues by producing wood and bark. As the stem grows in width, the **epidermis** breaks up and falls away.

In woody stems, bark replaces the epidermis as a protective covering. The bark is made up of a hard, dense tissue called cork. Cork insulates, waterproofs, and protects the stem. It also forms over wounds in the stem, like a scab forming over a cut in your skin. Only the innermost cells of the cork layer remain alive, because only they have access to the nutrients carried by the **phloem** and **xylem.** The older outer bark gradually wears away or splits apart and falls off as the stem grows wider.

Specialized stems

Some stems perform special functions. Bulbs and tubers, for example, are underground stems that can store large amounts of food. A bulb is actually a short stem surrounded by fleshy leaves. Onions are bulbs. Tubers are short and swollen and grow underground at the tip of the stems of plants such as potatoes.

Runners are specialized stems that are active in reproduction. Runners grow along the ground and produce new plants. Strawberries are examples of plants that spread by way of runners.

These are some varieties of onion (*Allium cepa*).

5 Looking at Leaves

The leaves are a plant's food factories. They are designed to gather energy from sunlight, then use this energy to make food (sugars). The sugars provide the plant with the energy and materials it needs to grow and to produce flowers and seeds.

Leaf parts

Most leaves have two main parts: the blade, or lamina, and the **petiole,** or leafstalk. The broad, flat part of the leaf is called the blade. This is where **photosynthesis** takes place.

The leaf blade is usually less than 0.039 inch (1 millimeter) thick. Such a thin, flat structure would fold if it were not strengthened in some way. A network of veins runs through the blade, which is made up of bundles of **phloem** and **xylem** cells that transport water and nutrients through the leaf. They are tougher and stronger than the **tissue** around them and prevent the leaf from collapsing or tearing.

The petiole is the stemlike part of the leaf. The veins run into the petiole, linking the leaf to the rest of the plant. In some plants, the petiole is quite thick. A stick of celery is a petiole. By contrast, plants such as grasses have no petioles.

In many plants, the petioles can bend and twist to move the leaf blades into the best position for gathering sunlight. You can actually see the leaves moving to follow the sun through the course of a day.

A typical leaf is broad and oval-shaped, but there are many other leaf shapes.

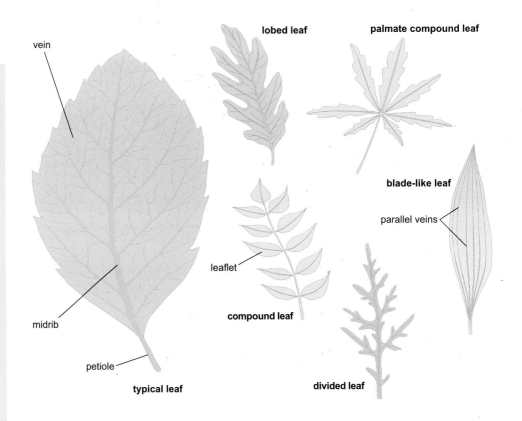

vein

lobed leaf

palmate compound leaf

blade-like leaf

parallel veins

leaflet

compound leaf

midrib

petiole

typical leaf

divided leaf

A tree's leaves spread in a wide canopy to catch the greatest amount of sunlight. The leaves are arranged in such a way as to overlap as little as possible.

As with the rest of the plant, the outer surface of the leaf is covered with a waxy **cuticle.** In plants that grow in bright sunlight, this cuticle is often extremely thick. This helps filter out strong light and keeps water loss to a minimum. The leaves may also have tiny hairlike projections to further reduce the effects of bright light. Plants that grow in the shade have a very thin cuticle, which allows in as much light as possible.

Leaf shapes

Leaves vary widely in their size and shape. A duckweed leaf may be no more than 0.039 inch (1 millimeter) across, whereas a water lily's leaves may measure 6.6 feet (2 meters) or more. Many leaves are oval, but others are shaped like arrowheads, feathers, hearts, spikes, and tubes. Leaves can be broadly divided into three groups according to their basic shape:

- Broad leaves are fairly wide and flat. This is the type of leaf that most **dicot** plants have, including oak trees, pea plants, and roses.
- Narrow leaves are long and slender. They are found on many **monocot** plants such as onions, lilies, and grasses such as barley, oats, wheat, and corn.
- Needle leaves are short, thick, and needlelike. They are typical of firs, pines, cedars, and other **conifers.**

Whatever their size or shape, most leaves are thin. This gives them a large surface area relative to their size for gathering sunlight.

Inside the Food Factory

The overall design of a leaf is adapted for **photosynthesis.** Under a microscope, it is possible to see how the inside of a leaf is designed to get the most out of available light.

Photosynthesizing cells

Most leaves are only a few layers of cells thick. This means that even the cells on the shady side of leaf get some light. The leaves of **dicots** have an upper and a lower surface. The upper side always faces the light, while the lower side is in shade. This is reflected in the leaf's structure.

The top surface of the leaf is a single layer of transparent **epidermis** cells. Immediately below this are the cells in which most photosynthesis takes place. These are slender, column-shaped cells, packed with **chloroplasts,** called the palisade **mesophyll.** Below this is a layer of larger, irregular-shaped cells called the spongy mesophyll. Large air spaces between the spongy mesophyll cells connect with the **stomata** on the leaf's surface. They allow carbon dioxide to reach the cells easily and quickly and oxygen to escape. The spongy mesophyll cells have fewer chloroplasts than palisade mesophyll cells, but they still photosynthesize.

In the leaves of **monocot** plants such as grasses, the palisade and spongy mesophyll are not organized into two layers. This is because the leaves grow vertically, so light falls on them from all directions.

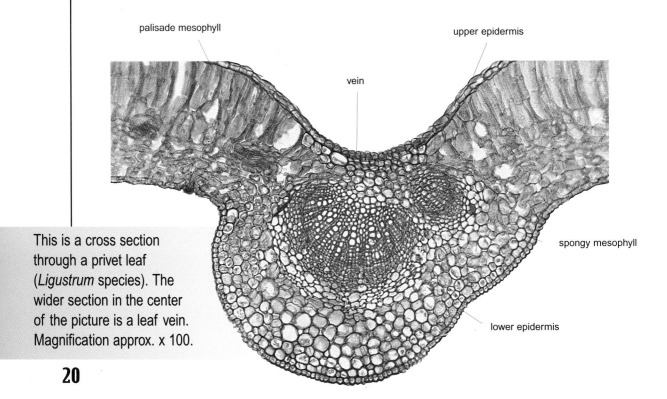

palisade mesophyll

upper epidermis

vein

spongy mesophyll

lower epidermis

This is a cross section through a privet leaf (*Ligustrum* species). The wider section in the center of the picture is a leaf vein. Magnification approx. x 100.

The colors of the autumn leaves on these trees in Maine range from brown to bright red to pale yellow.

Autumn colors

A leaf is green because of the **chlorophyll** it contains. During the growing season, a leaf becomes tougher as its cells develop thicker walls and it changes color from a bright green to a duller green. In the leaves of **deciduous** trees and shrubs, a corky layer of cells known as the **abscission zone** develops where the stalk of the leaf joins the stem. As autumn approaches, the shorter days and cooler nights cause a break down in the chlorophyll. As this happens, yellow and orange-red pigments in the leaf that had been hidden by the green of the chlorophyll are revealed. A group of red and purple pigments also forms in the dying leaf. After the chlorophyll breaks down, the leaf can no longer make food. The cells in the abscission zone separate or dissolve, and the leaf falls from the tree.

Photosynthesis

A leaf needs just three things for **photosynthesis:** carbon dioxide, water, and light. Using these ingredients, leaf cells make sugars. Carbon dioxide and water are the raw materials, and light provides the energy to join them together.

Leaves get their carbon dioxide from the air. Large numbers of **stomata** on their surface allow the air to enter. Because leaves are so thin, carbon dioxide does not have to travel far to reach the photosynthesizing cells. Water reaches the leaves from the roots, traveling up the stem and through the **petiole** in the **xylem** cells.

What happens in photosynthesis

When light strikes the **chloroplasts** in a leaf's cells, photosynthesis begins. Although there are a multitude of chemical reactions involved, photosynthesis can be summed up simply. There are two stages to the process. In the first, **chlorophyll** in the chloroplasts absorbs energy from sunlight. This energy is used to split water **molecules** into molecules of hydrogen and oxygen. This can be written as a simple equation (although in fact it is several different reactions):

$$\text{water} + \text{sunlight} \rightarrow \text{hydrogen} + \text{oxygen}$$
$$6H_2O + \text{sunlight} \rightarrow 12H + 3O_2$$

The hydrogen is then combined with carbon dioxide to produce a simple sugar. This second part of the process does not need light. The overall equation is like this:

$$\text{carbon dioxide} + \text{hydrogen} \rightarrow \text{glucose (sugar)} + \text{oxygen}$$
$$6CO_2 + 12H \rightarrow C_6H_{12}O_6 + 3O_2$$

The oxygen that is left over from both parts of the process leaves the plant through the stomata.

This is the overall equation for photosynthesis:

$$\text{carbon dioxide} + \text{water} + \text{sunlight} \rightarrow \text{glucose} + \text{oxygen}$$
$$6CO_2 + 6H_2O + \text{sunlight} \rightarrow C_6H_{12}O_6 + 6O_2$$

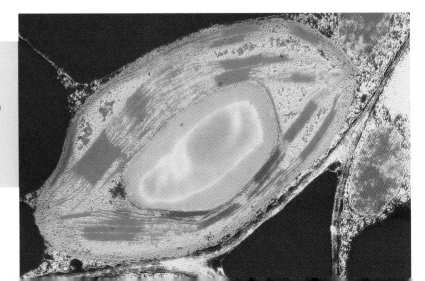

This is a false-color photo of a single chloroplast from a leaf cell taken with an electron microscope. The pink blob in the center is starch, which is the cell's main food store. Magnification approx x 20,000.

All from sugars

The glucose produced by photosynthesis is converted into a huge range of other materials. First, the glucose may be converted to other sugars, or to **carbohydrates,** such as starch and cellulose. Starch is the main energy-storing compound in plants, and many chloroplasts have large grains of starch in them that act as temporary energy stores. Cellulose, the main component of cell walls, is another important carbohydrate.

Lipids, the main components of plant membranes, are made from small molecules formed when sugar breaks down. **Proteins,** which control the many chemical reactions in the cell, cannot be made from sugars alone. Sugars contain only carbon, hydrogen, and oxygen. Proteins also contain nitrogen. So the plant needs nitrogen-containing **minerals,** which it absorbs through its roots. DNA, the cell's genetic material, also contains nitrogen and needs minerals absorbed through the roots.

Plant-eaters such as these giraffes get their energy directly from plants.

Light for life

Plants are at the beginning of an energy chain that supports nearly all life on Earth. They capture the energy of sunlight and turn it into food. Animals cannot make their own food as plants do. They rely on plants for their energy. Some animals eat plants to get food, and they in turn may be eaten by other animals. In this way, the energy of sunlight captured by plants is put to use in other living things.

Practically all of the energy available on Earth can be traced back to the sun. The fossil fuels, coal, oil, and natural gas, on which we depend so heavily for our energy needs, come from the remains of once-living organisms that trapped the light of the sun for photosynthesis many millions of years ago.

Sugar mountain

If the sugar made by all the world's plants was gathered together in one place, within three years the pile would be bigger than Mount Everest.

Leaf Specializations

Some leaves have other, special functions in addition to, or instead of, making food. Leaves also may provide protection, store food, offer support, and capture food.

Protection

Leaves that are specialized to protect the plant include bud scales, prickles, and spines. Bud scales protect the young, delicate **tissues** of growing buds. Prickles and spines are sharp leaf structures that discourage animals from eating the plant. For instance, prickles cover the leaves of the thistle and protect the plant from grazing animals. The clusters of sharp spines on many cacti are specialized leaves that not only offer protection but also decrease water loss in the dry desert. In a cactus, the green stem of the plant takes over the task of **photosynthesis.**

Storage

Most plants store food in their roots or stems. However, some plants have special leaves that hold extra food. Onion and tulip bulbs, for example, consist mainly of short, fat storage leaves called bulb scales. These leaves have no **chloroplasts** and cannot make food. Their role is to store food underground during the winter months, to provide the energy for new growth in the spring.

Many plants that grow in dry places have thick leaves that store water and look fleshy or swollen. These plants are called **succulents.** They include cactuses, agaves, and strange plants called living stones (*Lithops* species) that really look like stones. The **cuticle** of succulent plants has a waxy coating. These plants also have fewer **stomata** than other plants, and the stomata are sunk into pits in the cuticle to prevent water loss. The stomata never open during the day and, during periods of drought, they can remain closed altogether.

These sweet peas (*Lathyrus odoratus*) have some leaves adapted as tendrils for climbing and some leaves for photosynthesis.

Support

Many climbing plants have leaves that become slender, whiplike structures called tendrils. Tendrils wrap around twigs, wires, and other solid objects to help support the plant. For example, climbing garden peas have divided leaves in which the upper leaflets are threadlike tendrils. In one kind of sweet pea, the entire leaf blade becomes a tendril.

The floating leaves of a water lily have very little cuticle, as water conservation is not a problem. The **mesophyll** has huge air spaces, particularly around the **vascular** bundles. The air trapped there acts as a life preserver, helping to keep the leaf afloat. This specialized mesophyll is often called aerenchyma.

Food gathering

Plants such as the butterwort, pitcher plant, sundew, and Venus flytrap grow in wetlands, where the soil contains little nitrogen. To get the nitrogen they need, these plants "eat" insects. The leaves of these plants are specialized to attract, trap, and digest insects. The Venus flytrap, for example, has a leaf that is divided in two, with a hinge down the middle. If an insect lands on the leaf, the two sides suddenly close together, trapping it. Curved spines around the edge of the leaf keep the insect from escaping.

Bracts

Bracts are leaflike structures that grow just below the flowers of certain plants. Bracts are generally smaller and simpler in shape than the true leaves. Many members of the daisy family—such as daisies, marigolds, and sunflowers—have bracts that form a cup beneath the plant's cluster of flowers. In some plants, such as the poinsettia and flowering dogwood, the bracts are brightly colored and take on the role of attracting pollinating insects to the small-**petaled** flowers.

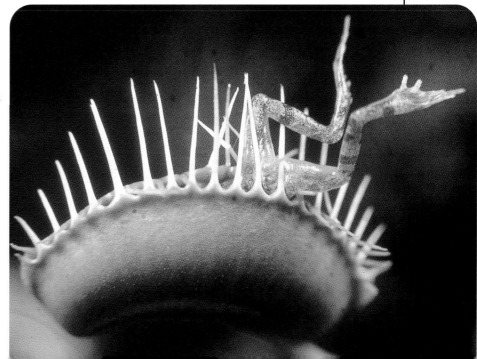

This Venus flytrap (*Dionaea* species) has caught an unusual victim—a young frog.

6 The Roots of Plants

When a plant seed begins to grow, the first thing to emerge is the primary root. Roots absorb water from the soil, along with nitrogen and **minerals,** such as calcium, magnesium, and phosphorus. These minerals are essential parts of important plant chemicals. Nitrogen, for instance, is found in **proteins** and DNA, while magnesium is a part of the **chlorophyll molecule.** The plant's **vascular system** carries water and minerals to all parts of the plant.

Hidden from sight beneath the ground, the **root system** can spread out over a tremendous area. This branching network of roots anchors the plant. More important, it provides a large surface area for absorbing materials from soil.

a) tap root system

primary root

lateral root

cortex

vascular tissue

epidermis

root hairs

area of meristem

root tip

root cap

c) enlarged root tip

b) fibrous root system

Root systems

The root systems of **dicot** and **monocot** plants grow in different ways. In dicot plants such as oak trees and carrots, the primary root grows downward into the soil, getting thicker as it grows. Lateral, or side, roots then begin to branch off from the primary root. A primary root together with its lateral branches forms a **taproot system.**

Monocot plant roots grow differently. The primary root of a monocot plant such as wheat or an orchid, is short-lived. Roots grow from the plant stem to replace the primary root and lateral roots branch from these new roots. The lateral roots are all about the same length and diameter. Roots formed in this way together make a **fibrous root system.**

The taproot system of a typical dicot plant (a), the fibrous roots of a typical monocot (b), and an enlarged view of a root tip (c).

Roots at work

At the tip of each root, a dome-shaped root cap protects the **meristem,** where cell division takes place. The root **epidermis** is where absorption of water and minerals from soil takes place.

Instead of using a bulb for food storage, a carrot (*Daucus carota*) uses an enlarged fleshy root.

Most of this absorption occurs near the tips of roots, through root hairs, which are fine extensions of the epidermal cells. A plant can have billions of root hairs. The hairs cling tightly to soil particles and provide a huge surface area through which water and minerals can be absorbed.

Cell membranes allow water, oxygen, and carbon dioxide to pass freely, but prevent other substances, such as salts, from passing through. When two solutions are separated by a cell membrane (or any other partially permeable membrane), water flows from the solution of lowest concentration toward the solution of highest concentration. This process is known as **osmosis.** In roots, the concentration of salts in the soil is less than in the root hairs, so water flows from the soil into the roots. The root hairs of a mature corn plant absorb more than three quarts (three liters) of water every day.

Some minerals are drawn passively into the roots along with the water. Others enter by simple **diffusion.** Plants can also take in minerals by active transport, which involves special proteins that actively pump minerals across the cell membrane. Active transport allows plants to bring minerals into the roots that would otherwise stay in the soil, but the root cells have to use energy to make it happen.

The endodermis

Water and minerals absorbed by the root hairs flow into the center of the root, where the transport **tissues** are situated. Surrounding the **vascular tissue** at the center of the root is a layer of cells called the endodermis. As with all cells, the membranes of endodermal cells let some substances into the cell but not others. They act as a border control for the plant, controlling the amount and type of materials that enter the plant's transport system from the soil.

7 A New Generation

The survival of any type of plant or animal depends on how successfully it can reproduce itself. Living things can reproduce in two ways: sexually and asexually. Many plants can reproduce by both methods.

In **sexual reproduction,** a male sex cell joins with a female sex cell (both called **gametes**) to produce a new plant. Both male and female gametes contain genes, which are the plant's hereditary material. A plant that is produced by sexual reproduction inherits genes from both parent plants. A plant produced by sexual reproduction is a unique individual, just as you are. It will share some characteristics with its parents, but will also have some characteristics that are different from both parents.

In **asexual reproduction,** there is only one parent. This parent plant divides into one or more parts, each of which becomes a new plant. The offspring are genetically identical to the parent plant and share its characteristics. They are clones.

This is a grove of aspen trees (*Populus tremula*) in the Boulder Mountains of Utah. The roots of an aspen tree can give rise to new trees by asexual reproduction. This produces rows or groves of trees with the same genetic makeup (clones). One such grove, nicknamed Pando (meaning "I spread"), covers more than 200 acres (81 hectares).

Vegetative reproduction

One type of asexual reproduction that occurs in many plants is **vegetative reproduction.** It can occur in several ways.

Some plants form special food storage organs in the summer growing season. In the autumn, the above-ground part of the plant dies, but underground, the storage organ remains. The following year, a new plant grows from the storage organ. This type of storage organ is called a perennating organ, a word that simply means "to live over from season to season." Onion bulbs, carrots, and potatoes are all perennating organs. The perennating organ in a potato is a swollen underground stem, called a tuber. A single plant may produce many tubers, each of which can give rise to a new plant. So a perennating organ is not only a way of getting through the winter, it is also a type of vegetative reproduction.

Many plants can reproduce vegetatively in other ways. Strawberries, for instance, have thin stems called runners that grow horizontally along the ground. The runners send out roots that produce plantlets. These plantlets are actually part of the parent plant, but if they are separated from the parent plant they will form new plants. Ferns, irises, many kinds of grasses, some shrubs, and some species of trees can reproduce from underground stems. Vegetative reproduction makes some weed plants difficult to control. A dandelion, for example, will regrow new stems and leaves from a fragment of root left in the soil.

Cuttings, grafting, and layering

Gardeners and farmers can take advantage of a plant's ability to reproduce vegetatively. A cutting is a part of a plant, usually a stem, taken from a growing plant. When placed in water or moist soil, the cutting develops roots and eventually grows into a complete new plant.

Grafting also involves cuttings. The cutting is grafted, or attached, to another plant, called the stock. The stock provides the **root system** and lower part of the new plant. The cutting forms the upper part. Farmers use grafting to grow fruit trees such as apple trees. Cuttings from trees that produce the desired variety of apples are grafted onto apple trees that have strong root systems.

Layering is a way of growing roots for a new plant. In mound layering, soil is piled up around a low-growing branch. This stimulates the growth of roots from the branch. The branch is then cut off from the parent plant and planted separately. In air layering, a cut is made about halfway through a branch. Moss is placed in the cut to keep it moist, and the branch is wrapped in a waterproof covering. New roots form in the area of the cut, and the branch can be cut off and planted.

Sexual Reproduction

Flowers are the reproductive organs of flowering plants. They develop from buds along the stem of a plant. Some plants have only a single flower, while others grow many large clusters of them. Plants such as dandelions and daisies have many tiny flowers that form a single, flowerlike head.

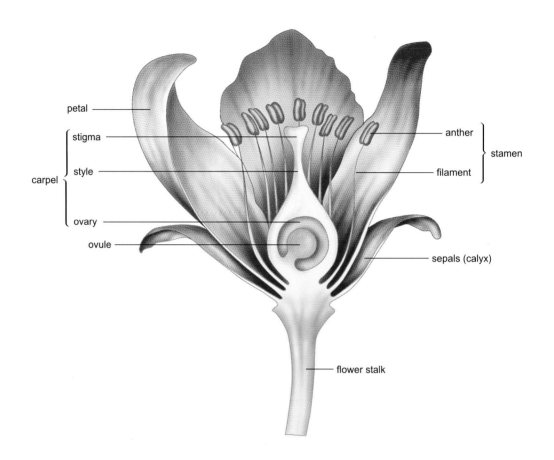

A cutaway view of a typical flower shows the main reproductive structures.

Most flowers have four main parts:
- The calyx is made up of small, usually green, leaflike structures called **sepals,** which protect the bud of a young flower.
- The corolla—the **petals** of the flower—is usually the largest, most colorful part of the flower.
- The **stamens** are the male reproductive parts of the flower.
- The **carpel** is the female reproductive part of the flower.

Sepals and petals

Like leaves, sepals and petals are made up of **ground tissue, vascular tissue,** and **epidermis.** Some of the epidermal cells in a petal produce fragrant oils. This is what gives many flowers a pleasing smell. Cells of the ground tissue frequently contain pigments that give the petals their bright colors. There also may be small, light reflecting crystals. This show of sight and smell, pleasant though it may be, is not for our benefit. Its purpose is to attract pollinators.

Stamens and carpels

The stamens and carpels are found inside the sepals and the petals. In many flowers, the stamens and petals are joined together. Each stamen has a long narrow stalk called a filament, on the end of which there is an enlarged part called an **anther. Pollen** grains, each of which develops into two male sex cells, are produced in the anthers. Many flowers have a single, centrally positioned carpel. Others may have more, perhaps fused together in a compound structure. The carpels of most flowers have three main parts. At the top there is a flattened structure called the **stigma,** a sticky or hairy surface that captures the pollen grains. From the stigma, a slender tube called the **style** extends down to the rounded **ovary** at the base. The ovary contains one or more structures called **ovules.** Egg cells form in the ovules. In most species of flowering plant, only one egg cell develops in the ovule. When sperm cells fertilize the egg cells, seeds form.

There is a huge variety of garden flowers, but these are only a small part of the much wider variety of flowering plants.

Separate flowers

Some species of plants produce flowers that have both male and female parts. These are called perfect flowers. Flowers that contain only male or female parts are called imperfect flowers. In plants such as oaks, a single plant can have both male and female flowers. In other plants, such as the willow, male and female flowers are on separate plants.

Pollination and Fertilization

Every spring, the **anthers** of flowering plants release vast quantities of **pollen.** For fertilization to take place, a pollen grain must be transferred from the male to the female parts of the flower. This transfer is called **pollination.** Some plants pollinate their own flowers. Pollen from the anther reaches the **carpel** on the same flower, or a carpel of another flower on the same plant. This process is called self-pollination. Other plants need to have pollen from another plant of the same species for fertilization to take place. This is called cross-pollination and requires the services of a pollinator.

A pollinator is anything that transfers pollen from the male reproductive parts of one flower to the female reproductive parts of another. Pollen grains can be carried from flower to flower by wind or water. Plants that are wind-pollinated have long anthers that produce huge amounts of pollen. Pollen is very light and can be carried for long distances on the wind. Wind-pollinated flowers are often small and inconspicuous with tiny **petals** or no petals at all. Many types of trees and grasses are wind pollinated. The water-dwelling ribbon weed releases tiny free-floating flowers called pollen boats. These drift across the surface of the water, and with luck they will bump into a female flower before they are swallowed by a fish!

A long-tongued bat (*Glossophaga sorciina*) feeding on a flower. The long tongue of this bat is ideal for reaching nectar. These bats are important pollinators.

Flowers and animals

Many cross-pollinated plants have large flowers, a sweet scent, and sweet nectar. These features attract birds, bats, and insects, such as ants, bees, beetles, butterflies, and moths. As these animals move from flower to flower in search of food, they carry pollen on their bodies.

One of the most important partnerships in nature is that between flowering plants and insects. Over the course of millions of years, plants and their pollinators have adapted to each other. In the beginning, all flowering plants would have been pollinated randomly by wind and water. However, when insects began to exploit the nutritious qualities of pollen, they accidentally made more accurate deliveries of pollen from flower to flower as they searched for food. The plant might have lost some pollen to the insects, but it gained a huge advantage over other plants because it successfully formed more seeds.

All around us we can see how plants and their pollinators have evolved together. Sweet-smelling flowers with light-colored petals, such as evening primrose and some tobacco plants, attract moths and bats in the evening. Red and yellow flowers of many tropical plants attract birds such as hummingbirds, which can see well but have no sense of smell. In Australia there are more than 1,000 flower species that are pollinated by birds. Then there are flowers such as *Rafflesia*, the largest flowers in the world, that smell like rotten meat and are pollinated by flies and beetles.

Fertilization

If a pollen grain successfully reaches a carpel, it starts to grow and develop into a tubular structure. This pollen tube grows down, carrying the sperm **nuclei** with it, through the **stigma** and the **style** to an **ovule** in the **ovary.** When the pollen tube reaches the ovule it ruptures, releasing its two nuclei. One nucleus unites with the nucleus of the egg cell and a new plant embryo begins to form. The second nucleus unites with another cell in the ovule and begins to develop into a food store for the new embryo. The plant embryo and its food store together form the seed.

This scanning electron microscope photograph shows pollen of an opium poppy (*Papaver somniferum*) clustered around part of the stigma. Pollen tubes can be seen growing from many of the grains. Magnification approx. x 650.

Seeds and Germination

Seeds vary greatly in size and shape. Some, such as those of the tobacco plant, are so small that 2,500 of them fit into a pod less than 0.78 inch (20 millimeters) long. At the other end of the scale, the seeds of one kind of coconut tree may weigh more than 20 pounds (9 kilograms). However, the size of the seed tells us little about the size of the plant that may grow from it. For example, giant sequoias grow from seeds that are less than 0.078 inch (2 millimeters) long.

The parts of a seed

A ripe pomegranate (*Punica granatum*) is tasty to eat, but its flesh is full of seeds. If the pomegranate is eaten, some seeds survive the journey through the gut and grow in the animal's droppings.

Seeds consist of three main parts: the seed coat, the food storage **tissue,** and the embryo. The seed coat protects the embryo, which contains all the parts needed to form a new plant.

In flowering plants, the food storage tissue is called the **endosperm.** It contains large amounts of energy storage substances, such as starch, plus smaller amounts of **proteins** and other nutrients.

The embryo is the part of the seed that will grow into a new plant. The embryo has either one or two **cotyledons,** or embryo leaves (one if it is a **monocot** and two if it is a **dicot**). These cotyledons absorb food from the food storage tissue.

Forming fruits

In all flowering plants, the seeds are enclosed by an **ovary.** As the seeds mature, the ovary develops into a fruit. Oranges, lemons, grapes, and other fruits that we eat are only one kind of plant fruit. Pea and bean pods are also fruits, as are dandelion's fluffy parachutes and the winged keys of trees such as maples.

The purpose of all these fruits is to protect the seeds until they are mature, then to spread the seeds as widely as possible. Some fruits burst as they dry out, catapulting the seeds away from the plant. Wings and parachutes help a seed to be carried by the wind. Many of the fruits that we eat have evolved to be eaten by animals. These seeds pass unharmed through the animal's digestive tract and are deposited perhaps miles from the parent plant in the animal's droppings.

Germination

The sprouting of a seed is called **germination.** Most seeds go through a period of inactivity called **dormancy** before they start to grow. In many parts of the world, this period lasts through the winter. Then, when spring arrives, the seeds start to germinate.

Seeds rely on environmental factors such as temperature, moisture, and oxygen levels in the soil to trigger germination. Most seeds, like most plants, grow best in a temperature between 64°F and 84°F (18°C and 29°C). They also need the right amount of moisture. Moisture softens the seed coat, allowing the growing parts to break through. If there is too much water, the seed may begin to rot. If there is too little water, germination may take place slowly or not at all.

Seeds also need oxygen for germination. To grow, the seed embryo needs energy. It gets this energy from its food store through **respiration.** Without oxygen, respiration is much less efficient and produces little energy.

From seed to seedling

The first structure to develop from the germinating seed is the primary root. Roots are important because seeds cannot store water and so must obtain it from the soil. Once the root starts to grow, a shoot begins to develop from the upper part of the seedling. At the tip of the growing shoot is the **plumule,** the bud that produces the first leaves. Until the seedling's first leaves begin to grow, its cotyledons are its main source of food. Once the seedling has developed roots and leaves, and has begun to produce its own food, the cotyledons wither away.

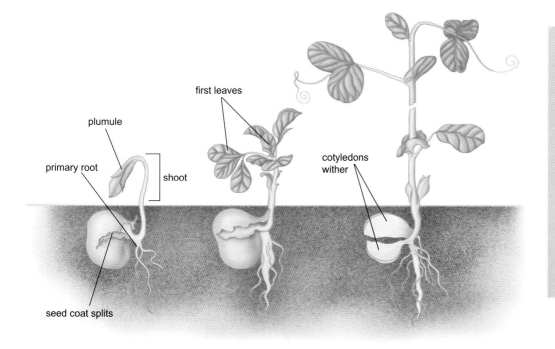

first leaves

plumule

primary root

shoot

cotyledons wither

seed coat splits

These are the stages in the growth and germination of a pea, which is a typical dicot seed.

Spores and Cones

Some plants do not produce flowers and seeds. The seeds of **gymnosperms** develop in scaly cones. In some plant life cycles, there are two distinct forms of the plant, occurring alternately. This is called alternation of generations.

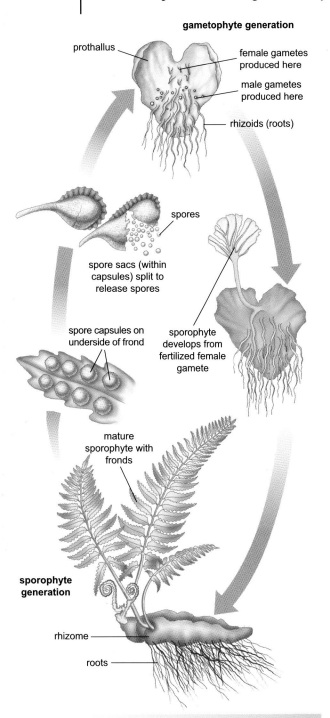

gametophyte generation

prothallus

female gametes produced here

male gametes produced here

rhizoids (roots)

spores

spore sacs (within capsules) split to release spores

spore capsules on underside of frond

sporophyte develops from fertilized female gamete

mature sporophyte with fronds

sporophyte generation

rhizome

roots

The two-stage life cycle of a fern shows alternation between gametophyte generation and sporophyte generation.

Mosses and ferns

Mosses and ferns are an example of alternation of generation. They go through a two-stage life cycle. During one stage, called the **gametophyte** generation, the plant produces its sex cells, or **gametes.** Mosses in the gametophyte stage are the small green plants you may be familiar with. In a clump of moss there are male and female gametophytes. At the tip of each gametophyte, eggs and sperm develop according to the sex of the plant. The male gametes have whiplike tails called flagella. They use these to move through the film of water on the plants to reach the eggs on the female gametophytes.

After fertilization the next stage of development is the **sporophyte** generation. During this stage, a long, thin stalk grows up from the fertilized gametophyte. At the top of the stalk, there is a podlike, spore-producing container. Inside there are thousands of single-celled spores. When these are released from the pod, they germinate and grow. Some will develop into male gametophytes and some into female gametophytes. The cycle begins again.

Ferns also go through a two-stage cycle. The fern plant we normally see is the sporophyte generation. If you look on the underside of a fern leaf you may see clusters of sporangia, or spore capsules. After the spores ripen, they fall from the sporangia to the ground, where they germinate and grow into tiny heart-shaped gametophytes. Fern gametophytes produce both male and female gametes. A new sporophyte will grow from the fertilized gametophyte, and the next generation begins.

Gymnosperms

Gymnosperms, such as the **conifers,** do not have flowers. Their reproductive parts are in cones. A conifer plant has two kinds of cones. The **pollen** cone, or male cone, is simpler in structure and is the smaller and softer of the two. Light, powdery pollen grains are produced here. The seed cone, or the female cone, is larger and harder than the male cone. Each of the scales that make up a seed cone has two **ovules** on its surface.

Pollen grains are released from the pollen cone and carried on the wind to the seed cone. Sticky surfaces near each ovule trap pollen grains, which then enter the ovule's pollen chamber. As in flowering plants, the pollen grain then begins to form a pollen tube. Once the pollen tube reaches the egg cell, it releases its two **nuclei.** One of the two nuclei fertilizes the egg. The other nucleus simply disintegrates; it does not help to form a food reserve for the new plant as in flowering plants. The fertilized egg develops into an embryo, and the ovule containing the embryo becomes a seed.

The word *gymnosperm* means "naked" or "uncovered seed." These plants have this name because the seeds are not enclosed inside fruits. Once the seeds are mature, the cone releases its seeds. The seeds fall to the ground and, if conditions are favorable, new plants begins to grow.

Ancient gymnosperms

Today there are only a few kinds of gymnosperms. These include about 750 conifer species, a handful of cycads (plants with a stout trunk and crown of large leaves), and a single species of ginkgo tree. In the Jurassic period, around 200 million years ago, ginkgoes, conifers, and cycads formed huge forests across the world. Cycads were so widespread that the Jurassic period, which we think of as the age of the dinosaurs, is sometimes called the age of the cycads. When the **angiosperms** appeared, about 140 million years ago, the gymnosperms began to decline.

These are male and female cones of the bristlecone pine (*Pinus aristata*). The small, red cone on the right is the male cone; the large cone in the center is the female seed cone.

8 Fungi

The fungi are often mistakenly thought of as plants. But they are very different and have their own place in the living world: the fungi kingdom. Fungi include familiar species such as mushrooms, puffballs, and toadstools, as well as less obvious organisms such as yeasts and the molds that grow on stale food.

Looking at fungi

Unlike plants, fungi do not **photosynthesize** and cannot make their own food. Instead, like animals, they must get their nutrients from organic matter. Like all other living things except bacteria, fungi are **eukaryotes.**

Fungal cells have walls, like plant cells. This cell wall may contain cellulose, as in plant cell walls. But most fungal cell walls also contain chitin, a substance similar to cellulose that is also found in the hard outer skeletons of insects and spiders. This difference between fungal and plant cell walls allows fungi to play an important role in the living world.

Although there are some single-celled species (for instance, yeasts), most fungi are made up of cells that form thin, microscopic filaments called **hyphae.** A network of hyphae together forms a **mycelium,** the body of a fungus. The hyphae form a complex system of microscopic tubes lined with **cytoplasm.**

World's biggest fungus?

The mycelium of a single fungus in a forest in Michigan was found to extend through 538,200 square feet (150,000 square meters) of soil. It is estimated to weigh more than two average-sized African elephants.

Parchment mushrooms (*Stereum complicatum*) are a kind of bracket fungus. They have enzymes that enable them to obtain nutrients from living and dead wood.

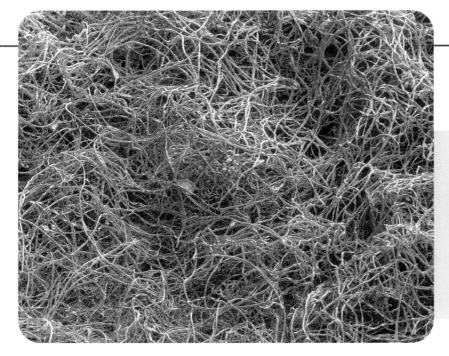

This photo, taken with a scanning electron microscope, shows part of the mycelium of three *Drechslera* fungus species. The mass of threads that fill the picture are hyphae. This type of fungus grows on grass and cereals.

Fungal feeding

Some fungi feed on dead and decaying animals or plants. Others are parasites that feed on living plants or animals, including humans. Growing fungal hyphae produce **enzymes** that digest their food outside the body of the fungus. The fungus then absorbs the products of this digestion. Fungi that attack plants can produce enzymes that break down the cellulose cell walls of plant cells, but have no effect on the chitin walls of fungus.

Decomposers

Most fungi live on dead organic matter. Alongside the bacteria, these fungi are the living world's clean-up squad. By breaking down once-living material, the fungi release valuable raw materials, such as carbon dioxide, nitrogen, and other elements for recycling through the ecosystem. The nutrients released by these fungal decomposers return to the soil, where they can be taken up once more by plants. Many fungi also help to maintain soil structure. They produce gluelike substances that bind soil particles together. This creates pores in the soil that allow air and water to filter through and reach plant roots.

Because fungi are such active decomposers, it is all too easy to see them as agents of destruction. For example, dry rot, which is a type of fungus, can cause a great deal of damage to buildings. However, in its natural setting, the fungus plays a vital role in recycling dead trees in forests. The cellulose-destroying enzymes allow fungi to attack all kinds of plant material, even wood. Without them and other decomposers, there would soon be no space for new trees to grow.

Spores, More Spores, and Buds

The mushrooms and toadstools you see growing from the ground are not whole fungi. They are simply the above-ground parts of a **mycelium** buried in the soil that could stretch far beneath your feet. Mushrooms are the reproductive organs of a fungus.

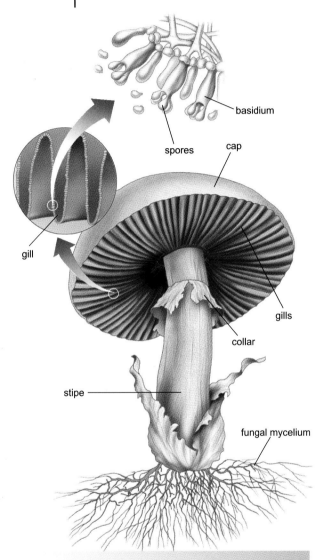

A portion of the fungal **hyphae** weave together to form the mushroom that we see. Look at one closely and you will see that it has a stipe, or stalk, and a cap. Underneath, the cap is lined with gills, fine sheets of **tissue.** The fungus produces spores on the gills. Each spore contains a little **cytoplasm** and one or more **nuclei,** depending on the species of fungus. These spores are the reproductive cells of the fungus. Fungi reproduce by producing spores in vast numbers.

Warning: poison!

A few kinds of fungi found in the wild are safe to eat. But many more fungi may make you sick, and some are highly poisonous. It is very difficult to tell poisonous and nonpoisonous species apart. Never eat a wild fungus unless it is identified as safe by an expert.

basidium

spores

cap

gill

gills

collar

stipe

fungal mycelium

The fruiting body of a typical mushroom shows the spore-carrying structures on the gills.

Mold fungi, such as those you can see growing on moldy bread, do not produce mushrooms. Instead they send up vertical hyphae from the mycelium. Chains of spores are produced at the tips of these hyphae, giving the fungus a blue-green, powdery appearance.

Spores are so tiny that they can be carried great distances by air currents. Once the spores land they will germinate, if they are on a suitable source of food. A melon-sized giant puffball can produce 700 million spores. Each one can potentially give rise to a new mycelium.

Spore prints

Spore color is an important indicator in identifying a fungus. One way to find the color of a mushroom's spores is to take a spore print. Making spore prints is easy. Simply cut the stem off a mushroom and place half the cap on a piece of white paper and half on black paper. (The two colors of paper are needed in case the mushroom has white spores.) Put a bowl over the whole thing and leave it for several hours. Now remove the bowl and carefully pick up the cap. You should have a pattern of spores on the paper that exactly matches the gill pattern of the mushroom. Mushrooms that look similar can have very different spore colors, so this is helpful for identifying a species.

Budding

Yeast differs from most fungi in that it consists of single cells. It does not reproduce by means of spores, but by budding. After the cell has reached a certain size, it produces a small growth, or bud. The bud gets larger and larger and eventually breaks away from the parent cell to become a new cell in its own right. On occasion the new cell will start to bud itself before it has separated. This can result in a small chain of linked cells.

Sexual reproduction

Spore production and budding are both forms of **asexual reproduction.** However, most fungi also have a **sexual** means of reproduction at some point in their life cycle. If the hyphae of two fungi of the same species meet, they may fuse together. This can result in fungal cells that have a nucleus from each of two different fungi. From this, a new mycelium forms that has genetic material from each fungi.

These apples have been infested with brown rot fungus (*Sclerotina fructigena*). In winter these fungi form a structure called a sclerotium, which consists of a few fungal cells inside a tough outer coating. The sclerotium can survive extreme conditions, then produce a new fruiting body in the spring.

Fungal Partnerships

Fungi have important relationships with plants, providing them with vital nutrients in the soil. But some fungi live in much closer partnerships with many other organisms.

These are lichens (*Eumycota*) on rocks on the Arctic island of Svalbard. Lichens are among the few living things that can grow in these harsh conditions.

Two different organisms living together in a close association are said to form a **symbiotic** relationship. Plants and fungi can form such relationships. In some symbiotic relationships, one partner gets all the benefit and the other gets nothing, or is actually harmed. The fungal parasites that destroy plant cells are examples of this kind of symbiosis. Some fungi such as blight and mildew are serious plant pests that cause a great deal of damage to crops. There are also a few parasitic fungi that cause diseases in animals.

Mycorrhizal fungi

Symbiotic partnerships do not have to be destructive. They may be of benefit to both partners. This type of relationship is called **mutualism.** One of the most important fungal relationships is a form of mutualism where plant roots and fungi grow together to form a **mycorrhiza.**

Mycorrhiza means "fungus root," and refers to the mutual relationship between fungi and plant roots, particularly young tree roots. Fungal **hyphae** radiate out through the soil, forming a velvety covering over the roots. Mycorrhizae are commonly found in temperate forests, particularly those of pine, beech, and birch. They help the trees to withstand seasonal differences in water availability and temperature. Some 5,000 species of fungus enter into these relationships.

In some forms of the relationship, the fungal hyphae actually penetrate the plant root cells. The fungus can absorb **minerals** from a larger volume of soil than the plant roots alone, and some of these minerals are passed on to the plant. In turn, the fungus absorbs sugars from the root cells. Effectively the fungi increases the roots' contact with the soil by 100 to 1,000 times.

Without the benefit of the fungi, plants do not grow as efficiently. About 80 percent of all **vascular plants** form such a relationship, although fewer than 200 species of fungus are known to be involved.

Lichens

Lichens are usually found where conditions are too harsh for most organisms. They colonize gateposts and walls, sun-baked rocks, and frozen mountaintops. They are excellent examples of mutualism. In a lichen, a fungus is entwined with **photosynthesizing** microorganisms called algae. Both grow and reproduce together.

A lichen may be leaflike, flattened, or erect depending on the species involved. The fungus is almost always the largest component of the lichen. The fungus benefits by absorbing a supply of nutrients from the algae. The algae, in turn, benefit from the shelter they get inside the lichen.

Vital to the earth

Fungi are not an obvious part of the environment, like plants. But fungal decomposers are important to the soil, and many plants (including whole forests of trees) depend on mycorrhizal fungi for vital nutrients. Unfortunately, many kinds of fungus are sensitive to pollution in the air, and the numbers and variety of fungi are declining.

If pollution deprives trees of their fungal partners, it will damage many forests, perhaps beyond repair. The loss of fungal decomposers would seriously affect the richness of the soil. The lives of animals, plants, and fungi are interconnected. Damage to one affects us all.

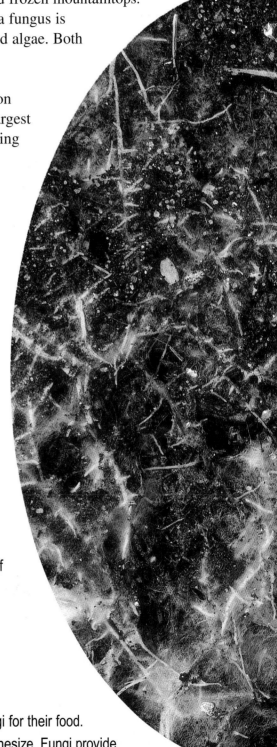

The roots of a lime tree (*Tilia vulgaris*) are covered with a white, fluffy covering of mycorrhizal fungus. The hyphae of this type of fungus do not penetrate the root cells. Magnification approx. x 500.

Orchids and fungi

Some types of orchids rely completely on mycorrhizal fungi for their food. These orchids have no **chlorophyll** and cannot photosynthesize. Fungi provide them with all the nutrients they need. Some fungi actually take nutrients from green plants and pass them on to nonphotosynthetic ones.

43

The Plant Kingdom

The plant kingdom consists of multicelled **eukaryotes.** Nearly all of them are **photosynthetic,** although there are a few that are parasites. There are nonvascular as well as vascular species. The vascular species predominate and have well-developed root and shoot systems. Nearly all are adapted for life on dry land, although a few are adapted to water environments. They mainly reproduce sexually, although **asexual reproduction** by vegetative propagation, for example, is also common.

Classification	Features	Examples
Phylum Charophyta	seedless, nonvascular	stoneworts
Phylum Bryophyta (bryophytes)	seedless, nonvascular	mosses, liverworts, hornworts
Phylum Psilophyta	seedless, vascular, no obvious roots or leaves	whisk ferns
Phylum Lycophyta (lycophytes)	seedless, vascular, leaves, roots, and stems	club mosses
Phylum Spenophyta	seedless, vascular, spore-producing	horsetails
Phylum Pterophyta	largest group of seedless, vascular plants	ferns
Phylum Cycadophyta	**gymnosperms,** vascular, naked seeds, simple cones, palm-shaped leaves	cycads
Phylum Gingkophyta	gymnosperms, seeds with fleshy outer layer	Ginkgo
Phylum Gnetophyta (gnetophytes)	gymnosperms	Epedra, Gnetum, Welwitschia
Phylum Coniferophyta	most common gymnosperms, cone-bearing with needlelike or scalelike leaves	
Family Pinaceae		pines, firs, spruces, hemlocks, larches, true cedars
Family Cupressaceae		junipers, cypresses
Family Taxodiaceae		redwoods, dawn redwood, bald cypress
Family Taxacea		yews
Phylum Anthophyta (angiosperms)	the flowering plants, largest group of vascular, seed-bearing plants; only organisms that have flowers and fruits.	
Class Dicotyledonae		
Dicotyledons (dicots)	two **cotyledons**	
Family Nymphaceae		water lilies
Family Papaveraceae		poppies
Family Brassicaceae		mustards, cabbages, radishes
Family Solanaceae		potatoes, eggplants, petunias
Family Salicaceae		willows, poplars
Family Rosaceae		roses, apples, almonds, strawberries
Family Fabaceae		peas, beans, lupins
Family Cactaceae		cacti
Family Cucurbitaceae		melons, cucumbers, squashes
Family Apiaceae		parsleys, carrots
Family Asteraceae		chrysanthemums, sunflowers, dandelions, lettuces
Class Monocotyledonae:		
Monocotyledons (monocots)	single cotyledon	
Family Liliaceae		lilies, hyacinths, tulips, onions, garlic
Family Iridaceae		irises, gladioli, crocuses
Family Orchidaceae		orchids
Family Arecaceae		date palms, coconut palms
Family Poaceae		grasses, bamboos, corn, wheat, sugarcane

Glossary

abscission zone area where part of a plant will separate from the main part of a plant

amyloplast structure found in the root cells of many plants, used for storing starch

angiosperm flowering plant.

anther part of plant containing **pollen**

asexual reproduction reproduction in which offspring arise from a single parent and are genetically identical to, or clones of, that parent

bryophyte a type of plant such as a moss or liverwort that has no internal transport system and requires the presence of free water to complete fertilization

carbohydrate chemical compound composed of carbon, hydrogen, and oxygen. Glucose is a simple carbohydrate

carpel female reproductive organ in a flowering plant, where pollen is received

cell membrane outer boundary of a cell. It controls what enters and leaves the cell.

chlorophyll light-capturing pigment found in plant cells that is involved in **photosynthesis;** it gives plants their green color.

chloroplast compartment found in plant cells that contains **chlorophyll.** This is where photosynthesis takes place.

chromoplast structure in a plant cell that contains pigments. Choloroplasts are a type of chromoplast.

collenchyma simple plant **tissue** that provides support

companion cell a type of cell that is specialized to help move materials into the other cells that will then move the mateerials throught the plant

conifer a type of plant, usually an evergreen tree or shrub with needlelike leaves

cotyledon seed leaf that provides a source of energy for the germinating seed

cuticle thin transparent covering of waxes and other materials on the outer cell walls of a plant's outer layer

cytoplasm all of the parts of a cell between the **nucleus** and the cell membrane

deciduous trees or shrubs that shed their leaves at the end of the growing season

dermal tissue tissue that covers and protects the outer surfaces of a plant.

dicotyledon (often shortened to **dicot**) a type of flowering plant that has two **cotyledons,** or seed leaves

differentiate develop and change to a more complex form

diffusion movement or mixing of substances as a result of the random motion of the molecules that make them up. The movement tends to be from regions of high concentration to regions of low concentration.

dormancy period during which the activity in an organism is greatly reduced. Dormant seeds help a plant to survive unfavorable conditions.

endosperm layer within a seed that acts as a food store for the plant embryo

enzyme one of a class of proteins that act as biological catalysts that greatly speed up the reactions that take place in cells

epidermis the outermost tissue layer of the plant

eukaryote cell that contains a nucleus and other compartments. All cells with the exception of bacteria are eukaryote cells.

fibrous root system branching network of roots growing from a young shoot.

gamete sex cell, such as an egg or sperm

gametophyte generation in the life cycle of a plant that produces the gametes

ground tissue tissue consisting of simple unspecialized cells that make up most of the bulk of a plant

guard cell one of two cells that lie next to each other in the surface of a plant's epidermis. When the guard cells swell with water, an opening forms between them through which carbon dioxide, oxygen, and water vapor can pass. When they lose water, the stoma closes.

gymnosperm plant, such as a conifer, that has its seeds exposed rather than protected inside the female reproductive organ as they are in flowering plants

hypha (plural **hyphae**) delicate filaments produced by fungi

lignin substance produced by plants to strengthen their tissues. It is the main constituent of wood.

lipids oils, fats, waxes, and other fatty substances found in living cells

Glossary

meristem regions of dividing cells in plants, such as shoot tips and root tips, where most growth takes place

mesophyll type of tissue where photosynthesis takes place. Palisade mesophyll cells are rodlike in shape and attached to the upper epidermis of the leaf. They contain a great many chloroplasts. Spongy mesophyll, below the palisade mesophyll, is less regular in shape and contains fewer chloroplasts.

metabolism sum total of the chemical reactions in a cell by which it acquires and uses energy for all activities that go on inside it

mineral simple chemical substance required by living organisms to function healthily. Plants get minerals from the soil. Animals get minerals in their food.

molecule particle made up of two or more atoms joined together

monocotyledon (often shortened to **monocot**) a type of flowering plant that has a single cotyledon, or seed leaf

mutualism type of symbiosis in which both species benefit from the relationship

mycelium network of hyphae that form a fungus

mycorrhiza a form of mutualism between the hyphae of a fungus and the roots of a plant

nucleus (plural **nuclei**) large structure in the center of a cell where its genetic material is held

organelle one of several different structures, surrounded by a membrane, found in eukaryote cells. The cell nucleus and plant chloroplasts are two types of organelles.

osmosis movement of water across a partially permeable membrane (such as a cell membrane) from a less concentrated solution to a more concentrated one

ovary enlarged base of one or more carpels in a flowering plant

ovule part of a plant **ovary** that develops into a seed when fertilized

parenchyma simple tissue made up of loosely packed thin-walled cells that makes up the bulk of a plant

petiole stalk that joins the leaf to the stem

petal often colorful part of a flower the function of which is to attract pollinators such as insects

phloem tissue forming part of a plant's transport system that carries sugars and other dissolved substances through the plant. It is made up of long cells called sieve cells that connect to form tubes and **companion cells** that help move substances into the tubes.

photosynthesis process by which green plants and some other organisms use the energy of sunlight to assemble glucose from carbon dioxide and water

plumule part of a seed that will develop into the plant shoot and carry the first true leave.

pollen tiny grains that carry the male sex cells of flowering plants

pollination process by which pollen is transferred from the male reproductive organs of a flower to the female reproductive organs of the same or a different flowe.

primary growth lengthening of shoots and roots as a plant grows

prokaryote cell that does not have its genetic material enclosed in a nucleus. All bacteria are prokaryotes.

protein one of a group of complex organic molecules that perform a variety of essential tasks in cells, including providing structure and acting as catalysts (enzymes) in chemical reactions

respiration process by which living things obtain oxygen from the environment to be used in the breakdown and release of energy from their food

root system parts of a plant that are below ground that take up water and help anchor the plant

sclerenchyma simple plant tissue with thick-walled cells that supports mature plant parts and protects seeds

secondary growth thickening of older stems and roots as a plant grows

sepal part of a plant formed from modified leaves that surrounds and protects a flower bud

sexual reproduction reproduction involving the formation of male and female sex cells or gametes and their joining together in the process of fertilization. The offspring inherit characteristics from both male and female parents and are unique individuals.

shoot system parts of a plant that are above ground, such as the leaves, stem, and flowers

sieve plate perforated cell wall allowing liquid to flow between sieve tube members

sieve tube members cells that join together to form the tubes that make up phloem

sporophyte generation in the life cycle of a plant that produces the spores

stamen male reproductive organ of a flowering plant, often consisting of a long stalk with an anther at the tip where pollen is formed

stigma sticky or hairy part of a carpel that captures pollen grain

stoma (plural **stomata**) gap or opening between guard cells on the surface of a plant's epidermis that open and close to control the movement into the plant of carbon dioxide and the movement out of oxygen and water vapor

style stalk that supports the stigma

succulent plant with thick, fleshy, water-storing tissues, adapted to live where water is scarce

symbiotic describes a relationship in which individuals of one species live alongside, in, or on members of another species for at least part of their life cycle

taproot system root system with a single main root, or taproot, from which other roots branch off. Most dicots have taproot systems.

tissue group of cells of the same type that work together to perform a particular task in a multicellular organism

tracheid one of the cell types that form part of a cell's transport system

transpiration loss of water from the above ground parts of a plant, especially from the leaves

vacuole fluid-filled cavity inside a cell that is surrounded by a membrane

vascular system system for transporting water and other material through a vascular plant

vascular tissue tissues that are part of plant's transport system through which water and other materials are transported through a vascular plant

vegetative reproduction form of asexual reproduction where new plants develop from multicellular structures that become detached from the parent plant

vessel member cell type that forms part of the **xylem** pipelines.

xylem tissue forming part of a plant's vascular system. Xylem cells form pipelines of interconnecting cells that carry water and dissolved substances through the plant. Xylem cells die when they reach maturity.

Further Reading

Attenborough, David. *The Private Life of Plants.* Princeton, N.J. : Princeton University Press, 1994.

Burnie, David. *Plant.* New York : Dorling Kindersley Eyewitness Guides, 1989.

Greenway, Theresa. *The Plant Kingdom: A Guide to Plant Classification & Biodiversity.* Austin, Tex. : Raintree Steck-Vaughn Publishers, 1999.

Wallace, Holly. *Life Processes: Cells and Systems.* Chicago: Heinemann Library, 2001.

Index